All About Weather

All About
Weather

A First Weather Book for Kids

Huda Harajli, MA

Illustrations by Jane Sanders

ROCKRIDGE
PRESS

To Hassan & Emily

What is
weather?

Step outside! What does the sky look like?
What does the air feel like?

The sky is blue!

The air feels warm!

That's the weather!

Weather changes all the time. It helps you decide how to spend the day.

Weather also changes throughout the year. These changes are called seasons. Many places on Earth have four seasons.

Spring

Summer

Fall

Winter

Spring weather
is warm and rainy.
It's when flowers start
to bloom.

Summer weather
is hot and sunny.

The air becomes cooler during fall and leaves change color.

Winter brings cold air and sometimes snow.

Near the middle of the Earth there are only two seasons: rainy and dry.

How many seasons happen where you live?

Thermometers tell us how hot or cold the air is. Higher numbers are hotter. Lower numbers are colder.

Rain freezes and turns into snow at 32° Fahrenheit (0° Celsius).

Is the sun
shining today?
Put on your
coolest shades!

What is the sun?

The sun causes the weather on Earth.
It provides light and warmth.

The sun is actually a star. It's more than 4.5 billion years old!

The sun can make the air outside feel hot.
Enjoy a sunny day by playing outside.

Remember to wear sunscreen!

You can ride your bike in the park or cool down at the beach.

Is it windy?
Quick! Catch your hat
before it flies away!

What is wind?

It is windy when the air around us moves.
We can't see wind, but we can feel it.

Enjoy a windy day by flying a kite
or blowing bubbles.

Windmills turn wind into electricity.

Are there fluffy shapes floating in the sky? It must be cloudy!

Cumulus clouds

Stratus clouds

Cirrus clouds

Nimbus clouds

What are clouds?

Warmth from the sun causes tiny water droplets to rise up into the sky. Those tiny droplets stick together and make clouds.

Enjoy a cloudy day by lying in the grass and watching the clouds change shape.

Clouds might look light and fluffy, but they're very heavy. One cloud can weigh more than one million pounds! That's the same as 100 elephants!

What shapes can you see? Name some.

Is it raining outside?
Grab your umbrella!

What is rain?

A drop of water spends nine days in the sky before falling back to earth.

Rain is water that falls from clouds. Rain has a very important job. It provides the water plants need to grow.

Enjoy a rainy day
by reading a book
near a window.

Listen to the
raindrops!

Is it snowy?
Lace up those boots!

What is snow?

When rain freezes, it's called snow.
Snow is icy and wet.

Snowflakes are different shapes, but they all have six sides.

Enjoy a snowy day by making snow angels.

Maybe try ice skating!

Is there a thunderstorm?

Let's move the picnic inside.

What is a thunderstorm?

If you can hear thunder and see lightning, that's a thunderstorm. Thunderstorms also bring rain and sometimes wind.

Thunderstorms form when cold air and warm air meet.

Enjoy a stormy day by playing games with your family.

Try counting the seconds between lightning and thunder. The higher you count, the farther away the storm is!

Stay safe inside and keep away from windows during a storm.

Look outside after a thunderstorm and you might see a rainbow. Rainbows form when light shines through raindrops.

The seven colors in a rainbow are red, orange, yellow, green, blue, indigo, and violet.

Let's get ready for the weather!

It's a cold, snowy day.
Point to what you will need.

It's raining!
Point to two items you will need.

The weather is sunny and bright!
Point to what you will need.

Look at the clothes each person is wearing. Use your weather knowledge to decide which season they're in: spring, summer, winter, or fall.

Now you're a
weather expert!

What is your favorite
type of weather?

About the Author

 Huda Harajli, MA is an author and elementary school teacher. She has undergraduate and graduate degrees in education from the University of Michigan. She loves helping young students learn about weather and weather safety. In her free time, she enjoys reading, writing books, and spending time with her family in Michigan's wonderful four seasons.

About the Illustrator

Jane Sanders illustrates children's books, magazines, newspapers, and even toy packaging. When Jane is not drawing, she loves to walk her poodle and French bulldog. She drew her French bulldog in this book!